BLACK
START

Black Start
Published by William Cook Publishing Ltd. 2023

ISBN 978-0-9934077-4-1

This special advance copy printed in the UK by CLOC Ltd. London
Soon to be available on Amazon in print and ebook.

Typesetting and cover design by www.bookstyle.co.uk with image
by Peter James Sampson on istockphoto.com

FOREWORD

My idea for *Black Start* arose from a conversation I had with Sir Peter Gershon, chairman of National Grid, about the vulnerability of wind and solar energy and the need to maintain an energy infrastructure independent of the vagaries of the weather. This reality has been borne out recently with the re-starting of Ratcliffe-on-Soar power station, which coincidentally was the model for my story. It strikes me as dangerously odd that the energy 'policies' of many western nations are dictated by political whim rather than common sense, in stark contrast to countries such as China, India and Russia, for whom the means of generation is irrelevant so long as they have sufficient electricity. *Black Start* is not just a story. It is a message and a warning to politicians and educationalists. Neglect to nurture your energy infrastructure and to train the engineers who design and operate it, and a new 'dark age' will soon be upon you – quite literally.

Sir Andrew Cook CBE, July 2023

BLACK START

START

by Andrew Cook

"Les?" The elderly lady, clad in dressing gown and holding a tray, looked anxiously at her husband, slumped in an armchair. "Les, it's the weather. Les?" Putting down the tray, she tapped him on the shoulder. "Les! It's the weather! You want to see it!"

Unusually for her age, the lady seemed to care about her appearance. Despite her attire of dressing gown and slippers, she still bore the signs of a much-faded beauty. Hair dyed brown, combed back into a girlish ponytail; a firm bearing, a trace of make-up. Even her slippers had kitten heels. Twenty years earlier she would have been a beauty. Now, though steadily losing the struggle against old age, she battled on. While there was still something to fight for, the lady would fight. For her husband; for herself; for her country, even; just as she fought her advancing years.

The old man stirred. "Put it on," he said, as if he had heard his wife all along but was reluctant to acknowledge this. The lady reached for the remote and pressed a button, "And now, the weather," came the announcer's voice, a map of the British Isles simultaneously showing on the wide screen. "No change," muttered the old man immediately. The map showed a cloudless sky over the entire country. "The high pressure which has blanketed the country and much of northern Europe for the past month shows no sign of changing." The announcer's chirpy blandness irritated the old man. "Temperatures tonight are forecast to reach minus eight over most of the country, and could fall as low as minus 20 on high ground... But here's some good news;" the announcer smiled; "The chill factor is zero. There is no wind."

"No wind, no gas, no power. How long will this madness last?" Muttering to himself, and oblivious to both the faux-cheeriness of the weather man on the television and the sound of the telephone ringing in the adjacent hallway, the old man had risen from his chair and was pacing frustrated around the room.

"Les, there's someone on the phone for you. Didn't give his name, but he said it was very important he spoke to you." People did not often telephone the old man. A phone call had been something of an occasion for him. He would usually take it with alacrity, until one unfortunate day when a well-spoken stranger had telephoned. The old man had been flattered when the caller informed him he had been selected as an experienced investor. In reality, the old man had seldom invested in anything other than savings certificates and premium bonds. Maybe this was what the caller meant. Certainly, he had never lost money. Twenty minutes later, the well-spoken caller had persuaded the old man to invest £10,000 of his hard-earned savings in a share. It was available only to 'selected investors', so the well-spoken caller had told him. It was expected to increase in value 'tenfold'.

A few weeks later, and having sent the £10,000 to the bank account he had carefully written down in accordance with the well-spoken caller's instructions, the old man had received a call from a lady who identified herself as an investigator from the Serious Fraud Office. Apologetically, she informed the old man that the 'wonder share' he had bought was in a non-existent company. Soothingly, if it is possible to sooth someone who has just lost £10,000, she told him he had been the victim of a so-called 'boiler room scam'. The old man was distraught, and vowed never to speak to strange callers again.

"It's one of those boiler-room people, no doubt! Tell them to go away!"

"Hello?" His wife spoke into the handset. "He says you're to go…" The elderly lady paused. "Yes, it is… Yes… Hold on a moment." The elderly lady turned towards her husband. "He says you're needed at the station. He knows of you, what you did. He sounds anxious; mentioned the weather… he insists he speaks to you."

The old man stirred. "The station, you say? Bring me the phone." The lady walked across the room with the cordless handset. Her husband had an old-fashioned mobile phone which he kept by him at all times, but this call had come in on the house landline. In any case, the transmitter masts on the mobile networks were frozen up most of the time. Only the landlines were still working reliably.

The old man took the handset from his wife. "Les Butterworth speaking." There was silence. The person at the other end of the line was obviously speaking. "Yes, I am." "Yes, possibly. It depends how it's been left. I need to go and see for myself." There was another pause. The old man was listening closely. "The army? Maybe. Hold on, I'll get a pen." The old man made frantic writing movements to his wife. She hurried over with pen and notepad. "Yes, go ahead. Major who? Yes, I've got that. And the number?" The old man wrote quickly on the pad. "I'll read that back," he said, a habit acquired from his days as a private pilot, when the read-back of air traffic clearances was *de rigueur* to avoid mistakes. The old man read out a name and two phone numbers. "Leave it with me. I'll know shortly." The old man pressed the 'end' button on the handset and threw it down on his chair. Without a word to his wife, he picked up his elderly Nokia mobile. "I'm sure I kept the number," he muttered to himself. "Yes, here it is." Reading the number from his mobile phone, the old man punched the numerals into the landline handset and pressed the call button. Seconds ticked by. "Come on, answer," the old man said to himself, pacing back and forth across the room. He was no longer an old man. There was, well, almost a spring in his step.

"Joe, glad you're in. It's Les. We're needed at the station. Can you meet me there? You've no fuel in your car? You're not sure? For God's sake, Joe! Look, do you have young McDonald's number? We'll need him. Try and get hold of him and tell him to meet me at the station. I'm on my way. I'll expect you if I see you." He ended the call. "Laura, I need the torches. I'm going to the station."

"But what if there's another power cut?" his wife responded, anxiously.

"There's certain to be. Go to your sister's. I could be some time." With that the old man grabbed his coat and scarf, picked up the two electric torches he kept in the hall, stuffed them in his coat pockets and walked out into the freezing dusk towards his car. Halfway across the driveway he hesitated. "I'd better take my tools," he said to himself, opening the car tailgate… "Good, they're still in the car." The old man had forgotten he had put his big toolbox in his car several weeks ago, 'just in case'. He opened the driver's door, got in and shut it, but before he could turn the key in the ignition, there was a banging on the window. It was his wife. Irritated, he wound down the window.

"Won't you need some help?" his wife said. Muffled up in woolly hat and scarves, the old man thought she looked quite attractive. He paused.

"Your car's still got some fuel in it?"

"Yes, I filled it up the other day, like you said: 'Keep it topped up while you can,' you said."

"Go round all your friends and collect up as many cans of fuel as you can. Tell them it's urgent. Tell them from me. They know what I used to do. Then bring them to the station. As many as you can! OK?"

The old man looked at his wife and smiled.

"OK," she smiled back. He wound up the window and turned the key in the ignition. With some reluctance, the engine started. It had not liked being left out in minus 20 degrees the past three days. Then came another bang on the window. It was his wife again. Exasperated, the old man lowered the window a crack, but before he could speak, his wife passed through her cigarette lighter.

"You might need this," she said. With a grunt of acknowledgement, the old man took the lighter, reversed out into the road and drove away, clouds of vapour pouring from the car's exhaust pipe in the freezing air of the January dusk.

As he drove, the old man pondered. He remembered his climatology since his grammar school days. This was the Polar High, a mass of cold air which normally resided over the polar ice cap. There had been atmospheric disturbances for the past 10 years and more: the media and the politicians called it 'global warming' and insisted it was man-made, caused by the burning of fossil fuels. The old man was not so sure. A keen alpinist in his younger days, he reminded any who cared to listen of the detritus left by Roman soldiers recently uncovered by receding glaciers. History showed that global temperatures had risen and fallen over the centuries irrespective of man's puny activities. But it suited the politicians to place the blame on the so-called polluters. Big power stations equated with big business. Coal was dirty and nuclear was dangerous, or so thought the ill-educated electorate, indoctrinated by the perverse ideology of the left-wing 'green' movement. Build windmills! They're nice and clean, they can be seen, and the wind is free! Oh yes, and we'll have some small, less visible gas-fired power stations as 'back-up'. That's the way to go! The old man had tried hard to explain the folly of this thinking to the various politicians who had occasionally visited his station, but he had been ignored. The windmills proliferated, and the gas plants too. In desperation, the old man had urged the need to expand the gas storage facilities. The country had capacity for only 20 days' supply. It needed four times this if it was ever to span the three months the more prescient climatologists believed a so-called 'cold snap' could last. His warnings had gone unheeded. There had been no wind now for three weeks: not just in Britain but right across northern Europe. The Dutch were keeping their gas for themselves.

The Russians had quadrupled the price of their gas and Germany was paying it. France's nuclear plants hummed away happily, the nation barely aware there was an energy crisis. Hydro power looked after the Swiss, the Austrians and Northern Italy. And Britain? Left out in the cold: literally. He had seen this crisis coming for many years. Irrespective of the alleged 'climate change', this crisis, this power shortage, was certainly man-made. Man could have seen it coming, and man could have taken avoiding action. English men, that is. Myopic, reckless, negligent, ignorant: whatever description you chose to apply to them, this crisis had been caused by politicians, aided and abetted by devious, influential and self-serving fellow travellers for whom facts were a mere inconvenience to be ignored if they went counter to their wishes. The old man gave an angry shudder.

Night had fallen, and the dark outline of the power station loomed ahead. He pulled up at the gatehouse. The gateman lurched out to greet him.

"Charlie! Thank God you're still here. Have you kept them turning?" The gateman was gaunt and bedraggled. Hair straggly, several days of stubble, dirty fingernails. Leaving the engine running, the old man got out of his car, walked across to the gateman's cabin, and entered. It was a slum. Filthy unwashed tea mugs scattered around, chocolate-bar wrappings, leftover pieces of pizza. Neglect leads to neglect. The man had been neglected ever since the power station had shut. Unremembered, unwanted, abandoned. A loner left alone. And when a loner is left alone, with no-one caring, he ceases to care himself. A decline sets in, most often accompanied by addiction to drink or drugs. Charlie had never done drugs in his life, but he loved the bottle and the evidence showed: heaps of empty bottles: wine bottles, whisky bottles, beer bottles, strewn around the cabin. A dustbin overflowed with them.

"What are all these bottles, Charlie?"

"I like a drink," the gateman muttered.

"Yes," the old man murmured under his breath. "So do I, and there but for the grace of God…"

"But I've kept them turning, just as you always said I must. They've never stopped." The gateman brightened up. "Never stopped! Had to use the standby much of the time, when the current failed, and there's little fuel left. But they're still turning: they're on the batteries right now."

"They'd have to be," the old man thought to himself. "There's no power."

The old man had instilled into his staff, including the gatemen, the absolute necessity of keeping the great turbine rotors slowly revolving. An auxiliary electric motor was attached to each for this purpose. If the rotors ever stopped, they would sink in their bearings. The amount would be microscopic, but enough to throw them out of balance, and the effect of this when revolving at the normal speed of 3,000 rpm would be to set up a vibration sufficient to cause the entire turbine rotor to shake itself from its mountings, burst through its casings and fly to pieces. A 150 ton mass of cylindrical metal with 600 pounds per square inch of steam behind it, flailing round at 3,000 rpm, could destroy the entire power station and even the nearby village if the monster escaped its confines.

"You still have the walkie-talkies? They should be in the store room." A look of anxiety flashed across the gateman's face.

"Er, I'm not sure," he answered nervously. The old man moved towards a door. The door led into the gatehouse store room. "Don't go in there sir!" exclaimed the gateman. But it was too late. The old man opened the door and flashed his torch inside. The beam revealed a chaos of empty bottles, a disorderly stack of beer cans and an assortment of magazines strewn around. The old man picked up one of the magazines, peered at it briefly and threw it down in disgust.

"Bloody hell, Charlie!"

"It gets lonely here sir, night after night," stammered the gateman shamefacedly. The old man shone his torch around the walls, halting the beam on a shelf screwed to the wall at waist height. On the shelf was a system of 'docks': receptacles of the sort one finds in cars to hold drinking mugs. The docks were square-shaped. They were the charging points for the walkie-talkie handsets used by the power station staff to communicate from remote points of the station in the days before mobile phones. They were all empty.

"Where are the walkie-talkies, Charlie?"

"Er…" The security guard muttered some meaningless noises. How was he to reveal that he had given the walkie-talkies to his grandchildren to play with?

"I… I loaned them out, sir."

"Loaned them out? To whom?"

"Er…" There was nothing for it but to tell the truth. "To my grandkids… to play with."

"You must get them back. Where do your grandchildren live?" The gateman stammered out a nearby address.

"Go and get them, Charlie. Now!"

"I've no fuel in my car, sir." Looking at the security guard's jalopy, it was a considerable surprise that his car was capable of mobility, fuel or no fuel. Just as it was that its owner was capable of driving it.

"Your landline working?" The old man gestured towards the telephone on the gateman's desk. Like everything else, it was filthy.

"I think so, sir." The old man picked up the receiver, wiped it with his handkerchief, and reading the number from his mobile phone, dialled.

"Joe, you not left yet? OK. Well, ask Greg to call at Charlie's grandkids' house and pick up the walkie-talkies." The old man turned towards the gateman. "What's the address, Charlie?" The gateman mumbled something incoherent. Annoyed, the old man handed the receiver to the gateman.

"You tell him." The gateman took the receiver and mumbled.

"Speak clearly, man!" the old man barked, sharply. The gateman repeated an address into the receiver.

Leaving the befuddled and confused gateman to contemplate the task of clearing up his rubbish, the old man strode out of the gatehouse towards the power station itself. Entering the turbine hall, he flashed his torch around the enormous dark cavern. The floor was littered with pigeon droppings, punctuated with cigarette ends and general rubbish. The old man fought back his outrage. In his day, smoking had been strictly forbidden throughout the station interior. Oil drums were scattered about, draped with dirty rags and left abandoned. The cold arctic moon cast its wan light through the translucent panels high up on the walls, silhouetting the curvaceous outlines of the great turbine generators, squatting on their foundations like giant hippopotami at rest. The old man headed for the steel steps that led up to the first platform. 'Number 1 Turbo-Alternator' read the sign on the side of the platform. Carefully shining his torch ahead to make sure there were no holes or loose steps to trap him, the old man climbed up to the platform. Moving along the side of the huge recumbent turbine generator, he paused momentarily to shine his torch on a brass plate. It was the maker's plate. On it, cast in the brass, he read 'C A Parsons and Co Ltd., Newcastle, England. Date: 1967. Serial Number:

153. Output: 500MW'. This machine, capable of producing three quarters of a million horsepower, was over half a century old. The factory which made it had closed decades ago. Now everything depended on him getting it going again.

The old man reached the end of the platform and clambered up onto the machine itself. Fumbling with a sort of flap, he lifted it and shone his torch into a hole beneath the flap. Was the shaft rotating? Screwing up his eyes and peering into the hole, he could just make it out. Yes, it seemed to be turning, very slowly. The old man shut the flap and clambered down to the platform.

"This is a 'black start'," the old man said out loud to himself. A 'black start' was the term used in the power generation industry for starting a power station from cold. "A black start without mains power to help us," the old man continued. For, without electricity, nothing could be done. Yes, this might be a steam-powered, coal-fired power station, but it still needed electricity to start it. Electricity for the coal mills and conveyors; electricity for the fans which blew the crushed coal to the burners. Electricity for the pumps which fed water to the boilers and condensers. There was no substitute. Electricity to start the bulldozer to push the coal into the feeders. And electricity to start the big auxiliary diesel generator, without which the station itself could not be started. Only electricity could do these things. There was no mains power, and there would be none unless and until he could start the station. Electricity from the batteries was all that stood between success and failure. No battery power, no black start.

Collecting the gateman on his way, the old man headed for the building which housed the station's auxiliary diesel generator. Like the rest of the station, the generator house was freezing cold and pitch dark. The old man flashed his torch around the walls. The battery bank seemed intact. The big diesel generator loomed, neglected, in the middle of the room.

"When did you last start it Charlie?" the old man asked.

"Not long ago sir." Another sheepish look crossed the gateman's face.

"How long?"

"Oh, a couple of weeks."

"A couple of weeks. You sure?"

"Well, it might have been a bit longer. It's a big job, starting this thing. Sometimes it won't start. And then there's the batteries…" the gateman paused.

"What about the batteries, Charlie?"

"They're often flat, sir."

"Often flat? Are they flat now?"

"I… I don't know, sir," the gateman stammered. "I've just been left alone… no-one to tell me what to do, no-one to care. Until you turned up two hours' ago, no-one's been here for weeks. My mate Kevin, on days, he's the same: comes and goes, keeps watch, does what he can, but you lose heart, sir. If no-one else cares, why should we?" The old man nodded. This was right, of course. If the bosses don't care, then why should the men? Such a disease seemed to have seized the whole country. There were a few old-school stalwarts still around: he was one of them, but they were a dying breed: literally. The decline was irreversible, or so it seemed.

"OK Charlie. So, how's the fuel?"

"I think it's OK, sir."

"You think? Let's have a look at the gauge." The fuel tank for the generator was outside the building but there was a gauge on the wall inside. The old man went over to the gauge and shone his torch against the glass.

"It's reading zero, Charlie."

"Is it?" The gateman's voice was querulous. He dared not admit that he'd siphoned off some of the diesel into his son-in-law's car. Several times. He did not like his son-in-law, but he loved his daughter and his grandchildren. And no-one would notice, would they? The station was shut. It would never be missed. Shame to waste it.

"Gone to the grandkids too, Charlie?" asked the old man, as if reading the gateman's mind? Then, dismissing the matter as irrelevant, "I think this gauge is frozen stuck. We'll have to sound the tank. Go and find a long stick, Charlie." Relieved, the gateman scurried off.

The old man went back outside the generator house to where a big square tank stood up against the wall. A ladder was fixed to its side. "I'll need my tools," the old man said to himself. He turned away and walked towards the gatehouse to find his car, returning in it a few minutes later. Opening the boot, he rummaged in his toolbox and brought out a large adjustable spanner. Holding the spanner in the fingers of one gloved hand so as to leave the palm free to grasp the ladder, he climbed laboriously up to the top of the tank.

"God, it's been a while since someone was up here," he exclaimed, looking at the rusted fuel cap, paint peeling off it. Taking the spanner in both hands, the old man managed to get a purchase on the groove in the filler cap. He heaved. Nothing moved. He heaved again. He heaved with such force that if the spanner had lost its purchase, the old man would have fallen off the tank onto the ground, a full 10 feet beneath him. That would, most probably, have been that. The old man knew the risks. As in much of his adult life, he weighed them up and made his decision. With experience, his decision-making capability had improved, as had his appreciation of the consequences of making the wrong decision. Here there was no choice. He had to get that filler cap off if he was ever to have a chance of starting the generator and keeping it going. There was no-one to help him: only the ruined gateman. Joe, the retired control room chief, hadn't arrived yet, and whether young Greg could be found, he didn't know. Only he, the old man, stood between success and failure: failure the consequences of which could be catastrophic for the country. That was no exaggeration.

The old man put all his weight against the spanner handle and heaved again. With a screech, the cap turned slightly. Again the old man heaved. With another screech, the cap turned some more. One more heave and it was turning more freely. Soon, he had it off.

"I've found a stick!" It was the gateman. The old man looked down from the top of the tank.

The gateman was brandishing what looked like a length of pipe.

"That should do, Charlie. Pass it up to me." The old man reached down and took the pipe. Pulling it up to the top of the tank, he plunged it into the open filler. 'Bump, bump' it went against the floor of the tank. "That sounds a bit hollow," the old man said to himself. He pulled the pipe back up out of the tank and shone his torch at the end. An oily sheen covered the bottom six inches of the pipe. Six inches of diesel. The old man did some rapid mental arithmetic. "The tank's three metres deep, that's 10 feet, and there's six inches of diesel left in it?" Not for the first time, he thought of his primary school teacher, Miss Baker, drumming mental arithmetic into her charges. Not physically drumming: Miss Baker never raised a hand to her pupils. Her tongue had been sufficient both to teach and to discipline, and mental arithmetic had been one of her fortes. The old man continued calculating in his head. The tank held 6,000 gallons. This would occupy say, eight feet of depth. Eight feet was 96 inches. There were six inches of diesel left. Six into 96 was 16. Sixteen into 6,000 was, say, 400. So there were 400 gallons left in the tank. The generator was 1,500 kilowatts, and it was powered by a 2,500 horsepower English

Electric diesel. "The same engine that powered the locos which hauled the coal trains," he thought to himself, his mind wandering. The English Electric diesel engine factory had closed years ago, and such British railway locomotives as still existed had German power plants. The old man involuntarily shook his head in a gesture of despair. He forced his brain back onto the matter in hand. "Four hundred gallons is 1,600 litres." Sometimes it was easier to do mental arithmetic in metric. The old man's still-agile brain could move seamlessly from metric to imperial and back to metric. "The engine would consume around 200 litres an hour at full power. Add a margin of 25 per cent, and assume 25 per cent of the remaining fuel to be contaminated by sludge at the bottom of the tank." He paused. "That makes four hours' running. Four hours. That should be long enough, but only just." He'd have to move fast.

The old man climbed down the ladder. The gateman followed him.

Back in the generator house, the old man shone his torch around. There, in the middle on its oil-stained pedestal, was the auxiliary power plant. The old man continued with his mental arithmetic. He tended to think in horsepower rather than kilowatts, and convert back arbitrarily at the end of the sum. One set of coal mills absorbed 500 horsepower; one fan set another 800; then there were the boiler feed pumps and the condenser circulation pumps: that was another 600 horsepower. That lot totalled 1,900 HP. Lastly 100 kilowatts, say 200 horsepower, for the auxiliaries: control circuits, lighting and so on. In total, he needed 2,100 horsepower, roughly 1,500 kilowatts, just to start one turbine generator set. Still, this was fair exchange if it delivered the expected 500,000 kilowatts in return.

The old man shivered. His shiver was only partly from cold. Yes, the cold was intense, and his hands were almost numb despite their gloves, but the real cause of the old man's shiver was nervous strain. The old man knew how, when he was under strain, a spontaneous shiver, or rather, a shudder, would sometimes pass through his body. It was his system's natural mechanism for releasing tension. It always worked. It cleared his mind and focused his judgement. The fuel would be half frozen. They'd have to get some heat into the generator house to have a chance of starting the diesel.

A car drew up and a tall figure climbed out. The figure walked towards the old man, hand outstretched.

"Hello Mr B. Joe called me and I came as soon as I could. Funny, I was half-expecting this." A look of pleasure on his face, the old man seized the outstretched hand and shook it vigorously, at the same time putting his left arm around the figure's waist in a part embrace.

"Good to see you, Greg! Thanks for coming!" The old man released his arm and stepped back. "So, how was the world of IT?"

"It paid well, but there was no interest in it. I felt I was just fiddling: 'fiddling while Rome burned', so to speak. And as your one-time 'apprentice', I couldn't really adapt to this virtual world." The apprentice looked up at the black outline of the station, cooling towers silhouetted against the faint light of the rising moon.

"This," he said, raising an arm in gesture, "This... this was real!"

"We've got to get it going again," the old man said, his hot breath steaming in the cold night.

"Well, I left it with plenty of coal, just like you'd always said I should." The apprentice smiled. "I did everything you taught me, Mr B, including maintaining a three-month coal stock. When they told me to run down the stocks, I ignored them. I knew we'd need it again one day. See over there?" The apprentice waved a hand in the direction of a vast, low mound, silhouetted even blacker against the blackness of the night. "I called it my 'coal wold'. There's over two million tons there, enough for 10 weeks at full power".

"Will the dozers still start?" the old man asked.

"They will if Charlie's done as I told him and kept the batteries charged up."

"Hmm, we'll see," the old man answered. "Charlie's gone to pot."

"Where's Joe?"

"Buggered off back to his wife. Mumbled something about needing some more warm clothes, and how she wouldn't like being left alone anyway, what with all these power cuts. Told me to pick up the walkie-talkies from Charlie's place, and he'd come along to the station as soon as he could." The apprentice paused. "Got some grief from Charlie's daughter. Had to be very firm with her. Didn't like the look of the bloke she had with her, either. Still, if it hadn't been for him, she might have given me even more trouble. 'Give 'em him!' he told her. Just like that. Nothing more. She looked scared, brought them to me in a cardboard box, and that was that. I think he understood what we had to do."

"You got jump leads?" asked the old man.

"Course I have, Mr B," said the apprentice. There was trace of indignation in his voice. The old man had trained the apprentice well, and he didn't like being treated like an idiot.

"Sorry. OK, see if you can get a dozer going while Charlie and me try to get this place warmed up."

"Will do, Mr B," answered the apprentice, getting back into his car.

The old man turned to the gateman.

"Come on Charlie, we need a couple of old oil drums: there's plenty scattered around. Some wood too. Broken pallets, that sort of thing. And rags; we'll need lots of rags. And your porn, Charlie. We can use that as kindling!" The gateman looked crest-fallen. The old man smiled and patted him on the shoulder.

Noises from the direction of the coal stock suggested that the apprentice had been successful in his mission to start a bulldozer. Half an hour later, he returned to the generator house. "I got a dozer started, Mr B. Shoved at least 500 tons into the feed hopper." "Fat lot of good that'll do us if we can't start this thing," answered the old man, gesturing at the generator. The three men were back in the generator house. Making and lighting a couple of oil drum braziers had not been difficult. The oily mixture of old rags, smashed pallets and hardcore pornographic magazines had ignited pretty easily, thanks to the cigarette lighter the old man had taken from his wife. The generator house was smoky, but at least it had warmed up.

"OK, let's give it a try." The old man pressed a red button on the starter panel. There was a click. Nothing happened. He pressed it again. Another click. "Shit! The battery bank's flat. We need to get the cars, and the jumpers. Greg?"

"My jump leads are only short, Mr B."

"Mine are longer. We'll have to do this in series. I'll drive my car up to the door, and you drive yours as close as you can get it to mine. Whatever you do, keep the engine running, we don't want to find we've shorted everything out. We're done for if that happens."

"Understood Mr B."

The old man and his apprentice drove their cars carefully up to the open doorway of the generator house. The old man nudged his into the doorway as far as it would go. The door was only wide enough to allow one car to protrude inside. The old man pulled the catch that released the bonnet. Meanwhile, the apprentice eased his car alongside, stopping just short of the generator house wall. He too released his bonnet catch and, getting out, lifted up the bonnet lid. Then he rummaged in the car boot and brought out two cables. "OK Greg," said the old man, "Connect those cables between the batteries, positive to positive, negative to negative, and don't get them mixed up."

"Right-ho Mr B, I've got it." The apprentice rolled his eyes.

The old man brought out two more cables from his own car boot. Longer ones, these, he proceeded to connect the clamps on the cable ends to the terminals of his own car's battery. The clutter of clamps on clamps made it difficult to maintain a good connection, with the clamps slipping on the terminals and on each other. When the old man was satisfied that the joints were clamped as best he could, he took the free ends of the cables and, nervously holding them well apart for, were they to touch, the batteries would be flat in a micro-second, he connected, firstly, the negative and then the positive to the end battery of the generator bank. Now to try again.

"OK Greg?" The old man called out.

"Ready when you are, Mr B!" came the reply.

"Both engines running?"

"Yes."

"Rev yours to 2,000 rpm."

"Will do."

"I'm trying it now." The old man went to the red button and pressed it. With a click, the starter motor engaged; and turned. 'Urgh Urgh Urgh Urghhh:' it nearly stopped. 'Urghhh…' a momentary pause, then… 'Vroom!' With a rattle and a roar, the diesel generator burst into life.

"It's started!" exclaimed the old man, needlessly. The generator slowed as the governor started to work and the oil circulated, before settling to a steady rumble. The old man reached for the main breaker switch and pulled it down. The diesel engine tone reduced slightly as it felt the load come on. "Probably for the first time in at least a couple of years," thought the old man. The light bulbs in the roof of the generator room flickered and lit. At last, he could see properly. 'Brrm… brrm… brrm… brrm… brrm,' rumbled the generator, thudding away steadily. The old man was tempted to sit back and leave it at that, but he checked himself. He had barely started. Yes, it was 'round one' to him, but many more rounds were still to come.

"Right, we have power," said the old man. "System's working here. Let's try to light a boiler." The men had left the generator house and walked across to the turbine hall. Entering the control room, now dimly lit with the power from the auxiliary generator, they confronted the long control desk with its array of switches, gauges and lights.

"I'll do this, Mr B," said the apprentice. "I was the last here. I shut it down, so I'll start it up." As he was speaking, the apprentice busied himself at the section labelled 'Number 1 Turbo-Alternator', running his fingers across the switches and buttons marked variously 'Coal Feed'; 'Mills'; 'Fan'; 'Feed Pumps' and 'Condenser'.

"I'll light the burners," said the old man. "Give me some of those oily rags, Charlie." A mood of confidence was in his tone. Was this over-confidence? The gateman passed across a bundle of rags soaked in diesel. The old man fumbled in a pocket and brought out the lighter his wife had given him. He flicked the wheel to check it worked. The flame ignited. "OK Charlie, you stay with Greg and help him. I'm going to the boiler house." Then, reaching into the box of walkie-talkies, "Are these charged up?" One after the other the old man picked up a walkie-talkie, pressed the 'on' button and seeing the red 'low battery' light, threw it down. Eventually he found two that displayed an orange light, meaning they still had some charge. The old man handed one to the apprentice and kept the other.

"Charlie, get the rest of them charged up," he said to the gateman, sternly. "Now! We'll need them later." With that, the old man walked out of the control room and headed towards the stairways ranged alongside the Number 1 boiler.

The power station boilers were of the high-pressure, water-tube design. Built by the once-famous British engineering company Babcock and Wilcox, they each comprised a huge rectangular steel box, some 200 feet high by 150 feet wide and the same deep, within which was a labyrinth of tubes arranged in the shape of an elongated lozenge. Above the tubes, and connected to them, was a barrel-shaped vessel. In the sides of the steel box were a large number of trumpet-shaped nozzles. A network of pipes led to these nozzles from outside the steel box. Coal dust, known as pulverised fuel or 'PF', was blown along these pipes to the nozzles, where it ignited. At the same time, water was pumped through the labyrinth of tubes. The heat from the burner nozzles turned the water to steam, which collected in the barrel-shaped vessel, known as the steam drum. From the steam drum, the steam passed to the turbines, causing them to rotate. In turn, the turbines drove the alternators, which generated the electricity. Lots of electricity. Five hundred megawatts per turbo-alternator: the equivalent of 750,000 horsepower.

Over-confidence is invariably followed by failure. Thus it was with the old man's initial attempts to light the burners. Repeatedly, he lit a piece of rag from his wife's lighter, reached with it through an inspection hatch and into the boiler cavity, waved it in front of a burner and watched it be blown out. There was something wrong with the system.

"Greg, can you hear me?" Pressing the 'transmit' button, the old man spoke calmly into the walkie-talkie. He released the button. The walkie-talkie crackled. There was no answer.

"Greg! Can you hear me?" The old man raised his voice, trying to conceal the anxiety in it. Again he released the 'transmit' button. The box crackled again. A tired voice spoke.

"Yes, hearing you OK. What's up?"

"There's not enough coal. Burners won't light. Are the mills both working?"

"I'll take a look," came the reply. "Stay switched on…"

Leaning against the gantry rails, the old man put his walkie-talkie on a convenient ledge, heaved a sigh, and pondered.

Each of the four huge boilers was supplied with coal from a pair of grinding mills. The coal was fed into the mills by a conveyor belt. Within the mills massive steel balls rotated around and around, grinding the coal to dust. A giant fan blew air into the mills, picking up the coal dust, the 'PF', and blowing it along pipes to the boiler burners. At its full power output of 500 megawatts, each boiler would require eight tons of coal per minute. Minute after minute, hour after hour, day after day, month after month, year after year: a giant maw, greedily gobbling coal, just so the lights stayed on, people could make tea, boil kettles, watch television, stay warm, go shopping... The list was endless. And what if the giant maw stopped, and the lights went out? A great, cold darkness would descend. A brief silence, a flickering of candles in the more prescient homes; then, soon after, the sound of smashing glass as the looters got to work; the occasional woman's scream, for uncertain but assuredly unwelcome reasons. The screech of tyres from cars driven with reckless abandon. The cries of men, outraged but helpless. For electricity is the great glue of civilisation. With light, there is safety. With darkness, there is menace. Take away religion, abandon the norms of Christian decency, hospitality and goodwill, and only electricity is left to maintain order.

"Greg? You hear me?" The old man refused to use the usual radio expression 'read', as in "Do you read me?" To him, 'read' meant 'read', not 'hear'. "Greg?" The apprentice's voice came back through the walkie-talkie.

"One of the mills seems to have jammed. I need to take a look inside."

"Don't forget to isolate it!" said the old man anxiously. Back in the 1970s, a man had been killed going inside a jammed mill without isolating it. It wasn't actually jammed. It had just tripped out when a piece of rock mixed up with the coal momentarily overloaded the motor. The rock was crushed, just like the coal, but the sudden additional load tripped out the mill. Then, while the man was still inside, someone else noticed the mill had stopped and pushed the 'reset' buttons to start it again. The screams had been audible even above the noise of the mills. When the fire brigade arrived to get the body out, one of the firemen was physically sick at the sight of the mutilated remains. What was left of the poor man was scooped up from inside the mill, put into plastic bin bags and taken to the mortuary.

After what seemed an age, the walkie-talkie crackled again.

"Mr B?"

"I'm here, Greg."

"One of the balls has broken. A big piece missing. It's jammed the system. I need a crowbar to get it out."

The old man sighed grimly. Back in the past, they'd used steel balls, made in a foundry in Sunderland. They never broke: they'd last for ever, but they were expensive. Then some bright spark looking for 'savings' decided to use balls made from a special iron. They were 30 per cent cheaper, but they were brittle and occasionally broke. The lesson went unlearned. The so-called cheaper, breakable balls weren't in fact cheaper to make, and the foundry making them went bust and closed. So the station was obliged to start using grinding balls from China. Yes, they were cheaper still, but even the most inexperienced engineer could see the defects. Chinese balls, even if they didn't break, lasted less than half the time of the old steel balls. The old man had tried to go back to steel. But the factory which had supplied them had closed long ago. Its site was now a shopping mall. It was China or nothing. The Chinese knew this, and, hey presto, the Chinese balls were now more expensive than the original steel ones had been.

The old man cursed: he cursed governments, he cursed managements: all this huge mess had been caused by their ignorance and incompetence. They were all the same: always looking for 'savings', always looking to be 'green' and all the time adding to cost, destroying dependability and all to no good purpose.

The walkie-talkie crackled. "Mr B, I've got it out; we're ready to re-start."

"OK Greg, let's try again." There was a pause, then a distant rising whine as the fans started turning. A slight hiss became a rushing noise as the air began to move. With a rumble, the mills started. The rumble deepened as the coal conveyor began discharging its burden into the mills. "Here comes the coal," the old man thought to himself. The rushing noise increased as the fans blew the air through the mills to the burners. The system was coming alive. It was time to try again.

The old man picked up his bundle of rags, still soaked with diesel, and fumbled for his wife's cigarette lighter. He flicked the wheel with his thumb. The lighter flared. Holding the lighter to the rags, he waited while the flame took. Then, keeping hold of the flaming rags, he pushed them through the burner inspection hatch. The air blast caught the rags, blowing them horizontally. The old man maintained his grip on the rags, waving them around, waiting for the coal to come.

"It's got to work this time," said the old man. "Come on, come on…"

He moved his hand as close as he could to the burner nozzle. A stream of coal dust started pouring out of the nozzle, like a jet of black water. The flaming rags caught the jet. There was a 'whoosh', and the black jet became a long, fierce orange flare. The rags, wrenched by the 'whoosh' from the old man's hand, disappeared into the cavernous void of the boiler. Further 'whoosh' noises followed, as the other burners in the immediate vicinity ignited from the heat of the first. Whoosh… whoosh… whoosh. With a sigh of relief, the old man shut the inspection hatch. Now he had to check the more distant burners. Wearily, he climbed the gantry steps. On each of the four levels there was a line of 16 burners. 64 burners in total and all had to be alight.

It took 20 minutes to get around all the burners. Most had self-ignited from the heat of the others. Of the 64, five had refused to light, requiring the old man once again to go through his flaming rag routine. In the end, all but one lit.

The old man suspected a blocked tube. But that didn't matter. With 63 burners together burning eight tons of coal a minute, they could raise steam. And with steam, they had power. Descending the staircases, the old man could already feel the heat soaking through the boiler casings. Carefully, he gripped the handrails. He couldn't risk injuring himself, now they were started.

As he made his way down the steps, the old man was reminded of Winston Churchill's famous statement after the Battle of El Alamein. "This is not the end: it is not the beginning of the end: but it is, just possibly, the end of the beginning." He'd got a boiler lit. If he could keep it lit and raise steam, then he could get a turbine going. With a turbine, they would have electricity, and with electricity, they could start the rest of the station.

Back in the control room, the old man studied the dials. Coal was flowing, water was circulating. No sign yet of any pressure.

"Where's Greg?" he asked. As if on cue, a figure entered the control room. The figure was black from head to foot. Black face, black hands, black hair: just the white of its eyes showing against the coal-black visage. Coal-black: it was exactly that, the blackness of the coal. The figure was coated thick with coal dust. "Shit," said the figure, slumping down in a chair. "Anyone got some water? Or some tea?"

"Charlie, go and make some tea," the old man

instructed the gateman. "We have juice for your kettle."

"I've no milk, Mr B."

"We'll have it black. Everything else is black!" The old man laughed at his witticism.

The control room lights flickered. The old man started, suddenly realising how much could still go wrong.

"How's the genny?" asked the apprentice, still slumped in his chair. Despite being served in one of the gateman's filthy mugs, the tea had revived him.

"I reckon we've about 40 gallons left."

"That's less than half an hour."

"I need an hour!" exclaimed the old man. "Where the hell's Laura?"

"Charlie!" The old man shouted into the walkie-talkie. The gateman was back in his gatehouse.

"Charlie? Any sign of my wife?" Not waiting for an answer, the old man walked out of the control room and into the freezing night. The sky was an inky blackness, relieved only by the stars and the hazy moon. No sign of street lights. No cars on the nearby motorway: most petrol stations had closed due to the power cuts, and motorists daren't get stranded in the arctic weather. People had literally frozen to death in their cars. A great, cold silence had descended on the country.

Looking up, the old man could see wisps of smoke coming from the tall power station chimney. "The first for years," he thought. "Well, it's a start."

Turning back towards the gate, he saw a car's headlights in the distance. It drew closer. The car halted at the gate. It was his wife.

"Come on, over here!" shouted the old man, waving. Moving forward rapidly, the car drew up alongside him. "How did you get on?" asked the old man.

"I got quite a bit," replied the lady. "I had some protests. Millie said she needed her car to go to bridge." The old man snorted.

"She had a huge stack of petrol cans up against the wall of her garage. Edgar, that's her husband, he's a terrible hoarder, and since these power cuts started he's been buying everything he could. All sorts of stuff, Millie said, but mostly candles, matches, fuel and booze: he's a dreadful drinker; that's why Millie escapes to the bridge club every evening she can. I told her there wouldn't be any more bridge if she didn't give me the cans…"

"Hang on!" Her husband interrupted. "Did you say petrol? We need diesel."

"I don't know: petrol, diesel, it's all the same to me."

"It isn't," muttered the old man. "Where are these cans?"

"I let down the back seat and filled the boot. There must be at least 20."

"What kind of car has Millie got?"

"Oh, I don't know: some German thing. It's got a 'D' at the end of the letters on the boot anyway. Does that mean it's diesel?" Her husband smiled briefly.

"Yes, it probably does." "Charlie;" he turned to the gateman, who had run up from his gatehouse, panting. "Show my wife to the generator house. Help her with the fuel cans. You'll have to climb up to the top of the tank. She'll pass the cans up to you." Allowing no time for protest, the old man turned away and walked back to the control room.

The old man looked at the pressure needle, inching its way around the dial. Normal working pressure was 600 psi. The needle had reached 450.

"We'll be down to fumes soon, Mr B," said the apprentice, coming through the open door from the generator house. "Got to give it a go."

"We'll do this by hand," said the old man. "Can't trust the actuators, and the valves are probably frozen stuck anyway. Come with me."

The old man and the apprentice went out of the control room into the turbine hall, still dimly lit by the emergency lighting. They climbed the steps to the gantry. At one end of the gantry, the end where the boilers were located, was an array of large control wheels, each operating a steam valve. Walking along the gantry, past the great low mounds of the turbine generator, the old man pondered on the ingenious British invention derived from an idea first sketched by Leonardo da Vinci and seized upon by Sir Charles Parsons, the British engineer. "Why have masses going backwards and forwards when you want something to turn round?" Sir Charles had posed this question. Ever since Newcomen 150 years earlier, steam had traditionally been used to move pistons back and forth in a cylinder. This was all very well if reciprocating motion was what was required, as in pumping water up from below ground, but if the end requirement was a turning movement, as with the wheels of a railway locomotive or the rotor of an electric generator, the crank mechanism required to convert reciprocation to rotation was inefficient and complex. Moreover, there was a limit to the extent reciprocating masses could move back and forth, stopping and starting at the end of each stroke. But if the motion was in rotary form from the beginning, the power output was theoretically limitless. Just as with the windmills of the Dutch polders, where the stronger the wind, and the bigger the windmill, the greater the power output, so with the steam turbine. The greater the pressure, the greater the power; and steam had another unique feature which wind lacked: its powers of expansion. By a system of turbine rotors of progressively greater diameter arranged end to end on a single shaft, all the energy in the steam could be absorbed right up to the point at which it turned back into water. Unlike on a steam railway locomotive, where the sharp puff-puff-puff of the barking exhaust was testimony to the retained energy being wasted, with the turbine every last kilowatt of power could be extracted and put to use. Silent and vibration-free, the power output of the steam turbine generator was limited only by the ability of the boilers to produce steam.

"We'll have to use the bypass valves, Greg. She'll not start just on the high pressure."

"OK, Mr B." The two men stood at the array of control wheels, each roughly a yard in diameter. When functioning properly, these valves were controlled by electric actuators. But with the valves half-seized, the actuators would not work. In any case, this part of the control circuitry had always been suspect. The old man had invariably preferred opening and shutting the valves manually. He liked the certainty it gave.

The control wheels, once polished and gleaming, were rusted and dirty.

"We'll open the bypass valves first. Ready? Give it a go…"

The apprentice heaved at a wheel. It didn't budge. He heaved again. Seeing him struggling, the old man moved to join him. Each took hold of a part of the control wheel rim.

"Together now… heave."

"It won't budge, Mr B."

"We've got to shift it, Greg." The old man's tone was calm. This was the crisis point. He had trained himself to be calm in a crisis.

"Wait a sec, Mr B. I've got some WD40 in my car." The apprentice hastened away. The old man leaned back on the turbine casing.

"Give it a good squirt, Greg." The apprentice had returned from his car with two aerosol cans of WD40 and a crowbar. He pointed and pressed, flooding the valve stem with the oily releasant.

"Right, no prisoners this time," said the old

man. Inserting the crowbar through the spokes of the wheel, both men heaved as one. The wheel moved. They heaved again. It moved some more.

"OK, that's freed it," said the old man. The apprentice grasped the wheel and turned it further. It still required considerable strength to move it, but at least it was turning. Another big squirt, and it turned freely.

Together, the two men managed to open both the first and the second low pressure bypass valves. "Right, now for the high pressure," said the old man. This was the crunch time; the critical moment. Open the high pressure steam valve and steam would start passing through the turbines, which should start to rotate. If the valve was jammed, that was the end of the matter. There was not enough fuel left to keep the auxiliary running long enough to attempt to light Number 2 boiler. This was it.

Taking no chances, the two men seized the control wheel of the high pressure valve, and heaved. The wheel spun freely: so freely, in fact that the men toppled over backwards. "Ouch!" shouted the apprentice, hitting the back of his head on the adjacent bearing pedestal. "Shit!" exclaimed the old man, picking himself up from the gantry onto which he had fallen, sprawling. He was oblivious to any pain or discomfort the tumble had caused him. He seemed unhurt. The old man rarely swore except in cases of extreme stress. "Shit!" he said again; then, looking at the wheel-less valve stem, "Christ Almighty, the wheel's come off its shaft!" Inexplicably, the high pressure control valve wheel had fractured at its hub and come off. The valve stem projected nakedly out from the valve body.

"Where's my big spanner?" The determination in the old man's voice was palpable. "Damn. Where did I leave it?" The old man was thinking about the spanner he'd used to unscrew the filler of the auxiliary generator fuel tank. Had he left it on top of the tank?

"Greg, I think I left it on top of the fuel tank."

"I'll go and look." The apprentice hastened down the steps from the gantry to the power station floor. Three minutes later he was back with the spanner. He handed the spanner to the old man. The old man took it and offered it up to the valve stalk, turning the adjustment screw until the jaws bit.

"No prisoners again!" said the old man, applying his weight to the spanner.

"You said that last time. This might help." The apprentice inclined his head downwards towards a scaffold pole he was holding. He slid the pole over the spanner handle. The extra length would double the leverage on the valve stem. "I'll hold the jaws on, you heave on the pole," said the old man. "Ready?"

"Yes."

"Heave!" The apprentice heaved, the valve stalk turned, the jaws bit, then slipped off.

"Just a sec while I put it back on." The old man fitted the spanner jaws back over the valve stalk. Again, the apprentice heaved. With a screech, the valve stem moved maybe one eighth of a turn. Again, the apprentice heaved, and as he heaved, the old man squirted the WD40 at the stuck valve spindle. This time the jaws stayed on. The stem turned some more. Six more times the old man and his apprentice repeated this process, heaving and squirting, the valve stem turning a little each time.

"Right, that's well over a full turn; let's take a look." Leaving the control valves, the old man strode to the far end of the gantry, climbed up to the rearmost bearing pedestal and opened the inspection flap. He peered in. Was the shaft turning faster?

"We need to open the valves up to full," the old man called along the gantry. The apprentice picked up the spanner. Heaving on the scaffold pole again and again, the two men slowly opened the valve. Presently, the spanner would turn no further.

"That's as far as it will go," said the old man, leaning back exhausted against the turbine casing. A low hum, just audible, began. The old man placed the palm of his hand flat against the turbine casing.

"It's starting!" he exclaimed. "Come on, let's get these other valves full open!"

The hum rose in tone and became louder. "Mr B, we'll have to shut the bypass valves," called out the apprentice. The bypass system admitted high pressure steam to all stages of the turbine and was needed to get it started, but for the system to work efficiently, the steam had to pass progressively through all the rotors end to end in sequence. Here was Boyle's law in operation, the steam expanding progressively as it cooled and the pressure dropped, all the time imparting its energy to the spinning rotors.

It took another 20 minutes for the two men to force open and shut the remaining valves, and a further 40 for the turbine to be running at full speed. The analogue revolutions indicator at the side of the generator read just short of 3,000.

By now, they were both exhausted.

"God, I could do with a cup of tea," said the apprentice.

"No time," said the old man. "We've got to get the power on. Even with Laura's cans, the standby will be out of fuel any time now." The two men hastened back to the control room.

The old man and his apprentice re-entered the control room, still only dimly lit by the emergency lighting. Both knew exactly what to do. They stood before the long control desk, examining the dense array of buttons, dials and switches. The control desk was divided into four separate sections, one for each turbine generator set. Standing in front of the section over which was the sign 'Number 1 turbo-alternator', they viewed the dials, lights and switches.

'Mills running': a green light.

'Fan running': another green light.

'Coal feed': a dial this time, indicating a flow rate in tons per minute.

'Feed pumps': two lights this time, both green.

'Condenser pumps': another green.

'Steam pressure': three dials here, marked 'high', 'intermediate' and 'low'. The 'high' needle was hovering on '600', meaning pounds per square inch.

'Turbine rotor speed': another dial, with the needle pointing to 3,000, meaning three thousand revolutions per minute.

Then there were the electrical gauges: volts, amps and the all-important megawatts. All needles pointed to zero. Beneath the electrical gauges was a large, prominently labelled switch. 'Main Breaker' read the label. Alongside the switch were the labels 'engaged' and 'disengaged'. The switch was in the 'disengaged' position.

"OK Greg, everything looks right; let's give it a go." The old man grasped the switch and pulled it towards him. From outside the control room came a 'crash' noise, rather as a car would make crunching into a hard object, like a wall. The lever sprang back to the 'disengaged' position. The old man looked concerned. "Funny," he said. "It shouldn't do that." He pressed a black button labelled 'reset'. "Let's try again." Again the old man pulled the main breaker switch towards him. Again came the 'crash' noise.

"Grid's in overload, Mr B," said the apprentice. "Breaker can't take it."

"Shit!" exclaimed the old man. "We'll have to frig it. Where are the Castell keys?"

"They should be in the cabinet over here," answered the apprentice, walking to the far end of the control room to where a grey metal cabinet was fixed to the wall. To their joint relief, the cabinet was not locked. Had it been, they would have forced it open. The apprentice opened the cabinet door. Inside were four hooks and on each hook there hung a grey square key, not unlike the keys with which old clocks would be wound up. The hooks were labelled 'Number 1', 'Number 2', 'Number 3' and 'Number 4'. The apprentice took the key off the 'Number 1' hook and, the old man following, went out of the control room and back into the turbine hall.

Set into the wall opposite the alternator end of Number 1 turbine generator was a door marked 'switch room'. The apprentice opened this door. Inside was a large enclosure caged off by a heavy wire mesh. Inside the cage was an array of cables and electrical equipment. Protruding through the cage mesh was a lever. The apprentice inserted the square key into a socket alongside the lever and turned it a half turn. This released a lock which prevented the lever from being moved. The apprentice then took hold of the lever and pushed it down.

"That's isolated it," he said. The act of pushing the lever fully down had both isolated the main breaker electrically and released the door of another grey box, this one situated at the bottom of the lever's range of movement. The apprentice opened the box and took out another square key.

This he inserted into the door of the cage and turned. The door swung open and the two men entered the cage enclosure.

The Castell locking system was an ingenious mechanical interlocking arrangement invented in the early years of electric power to ensure no-one could enter an electrical enclosure when the system was live. It was fool-proof. Later systems had been more complex and sophisticated, but they required electricity to work. The old man had no time for the later safety devices and considered the original Castell system to be unsurpassable.

The two men went up to a big rectangular casing in the middle of the cage. Opening the door of the casing revealed a complex contraption, much as one might imagine HG Wells' 'The Time Machine' to look. Strips of copper plate, known as 'bus bars', were bolted to the contraption. The contraption was the main circuit breaker and the bus bars conducted the electricity to and from it. The contraption's main mechanism included three huge coil springs the size of a heavyweight boxer's thigh, and each spring was kept in a state of partial compression by a washer the diameter of a dinner plate, capped with a nut like an octagonal saucepan.

"We're going to need your spanner again, Mr B," said the apprentice.

"Damn, where did we leave it?" exclaimed the old man.

"Back on the platform. I'll get it." Leaving the cage, the apprentice was soon back, carrying the big adjustable spanner over his shoulder.

"I'll do this, Mr B." The apprentice seemed to have a second wind. They were nearly there now. Success was within their grasp.

"Steady now," said the old man to the apprentice, mindful from recent experience of the dangers of over-confidence. Opening the jaws to their maximum, the apprentice fumblingly forced the spanner over the massive nut, and heaved. The spring compressed fractionally. Again and again the apprentice fumbled and heaved until all three springs were fully compressed.

"That's it, Mr B. It can't trip now."

"No, but it can burn out," thought the old man to himself. Then, out loud, "OK let's try again."

The two men returned to the control room.

"Here we go," said the old man, anxiously. He pulled the breaker switch towards him. There was a 'clunk' from outside the control room. The needles on the gauges swung wildly across their dials, and settled. To the steady whine of the turbine generator was added a fizzing noise; the noise of high-voltage current emerging from the alternator.

"It's held!" exclaimed the old man. "We're generating."

Just then the gateman entered the control room. "Diesel's misfiring," he said. "I think we're out of fuel."

"Doesn't matter, Charlie," said the apprentice. "It's working. We've got power. You can shut it down." The old man and the apprentice looked at the dials. The needles were all in the right places, with the all-important 'MW' megawatt gauge pointing at 500. They looked at each other and shook hands. For a moment, neither could speak. Then the old man cleared his throat and spontaneously hugged the apprentice. "Thank you, dear boy," he said. "It would've been impossible without you."

Another man entered the control room. The old man turned and looked at him.

"Joe, where the hell have you been?" The other man looked sheepish.

"I had to get some warm clothes and then the wife, she didn't want me to leave her, and by the time…" His voice tailed off.

"OK Joe, enough excuses."

"And I had to show these chaps the way to the station," said the man called Joe, as if attempting to redeem himself. "Found them stopped at the crossroads looking at a map." Three more men, all in army uniform, came into the control room. The first held out his hand towards the old man.

"Major Barnes, REME;" the army officer pronounced it 'reemee' the universal shorthand term for the Royal Electrical and Mechanical Engineers. "We've come to help you get it going." The officer paused, looking at the control panel. "But it seems you've managed without us." The soldier turned back to the old man and smiled. "What can we do?" he asked.

"Major, we've done the difficult bits. Number 1 is running. Power's flowing into the grid. Get 2, 3 and 4 going, and keep the whole show running for the next 12 hours till we come back. Joe will help you." The old man shot a scornful look at the man called Joe, who had once been his assistant before the apprentice replaced him. "Get your squaddies on the bulldozers quickly. This thing likes coal. Oh, and we had to frig the breakers to stop them tripping. The whole grid's running in overload. You'll have to reset them or they'll burn out sooner or later." The old man paused.

"How are the others doing?" he asked. The major knew the old man meant the other coal-fired power stations.

"Drax is started. Proved a pig, but once we went 100 per cent to coal instead of those useless wood pellets the green brigade forced it to convert to, we got a boiler lit and the rest followed. Cottam's running too. I'm not sure about the rest."

"Well, me and my mate;" the old man gestured towards the grimy apprentice, slumped exhausted in a control room chair: he sometimes liked to emphasise a bloke-ish camaraderie where a particularly challenging task had been completed by him with assistance from another: "Me and my mate, we're going home. Come on Greg, you'll fall asleep at the wheel. I'll give you a lift. Charlie, you come too. You deserve a rest." The exhausted apprentice dragged himself to his feet and the three men left the room.

Instead of heading straight to his car, the old man lingered for a moment in the turbine hall. The huge Number One Turbo-Alternator was humming away steadily. The hall now brightly lit from mains power, the mess and disarray which had disgusted the old man when first seen by the light of his torch several hours previously seemed less obtrusive. It wouldn't take long to clear it up. Looking at the great machine, the old man began to recite Kipling's 'McAndrew's Hymn'.

"They're all awa', true beat, full power, the clanging chorus goes; clear to the tunnel, where they sit, my purrin' dynamoes".

Well, it wasn't so much a clanging chorus the old man was witnessing, Kipling's ship's engineer being from the era of reciprocating engines, but the literary analogy was still close.

Dawn had broken when the old man got back to his house. Opening the front door, he was greeted by his wife. Despite the rigours of the night, she appeared clean and fresh, dressed in a silk night gown. "The electricity's back on. Let's have a cup of tea," she exclaimed, brightly.

"In the kitchen?" asked her husband.

"No, I think we'll have it upstairs."

"I need a shower," said the old man.

"They'll be time for that later. I want you as you are." The old man looked startled. "Oh, and have you still got my lighter? I'm going to need a cigarette!" A smile on her face, the lady took her husband by the hand and led him up to their bedroom.